discovermore

Exploring Physical Science

Gravity

Marie Harts

Britannica Educational Publishing

IN ASSOCIATION WITH

THE ROSEN PUBLISHING GROUP

Published in 2024 by The Rosen Publishing Group, Inc.
2544 Clinton Street, Buffalo, NY 14224

Editor: Shannon Harts
Book Design: Rachel Rising

Photo Credits: Cover Maridav/Shutterstock.com; (series background) Dai Yim/Shutterstock.com;p. 4 Mark Herreid/Shutterstock.com; p. 5 Orla/Shutterstock.com; p. 6 Sergey Novikov/Shutterstock.com; p. 7 Iryna Kalamurza/Shutterstock.com; p. 8 https://commons.wikimedia.org/wiki/File:Isaac_Newton,_English_School,_1715-20.jpg; p. 9 f11photo/Shutterstock.com; p. 10 Jurik Peter/Shutterstock.com; p.11 https://commons.wikimedia.org/wiki/File:Saturn_2019.png; p.12 https://commons.wikimedia.org/wiki/File:Albert_Einstein_1921_by_F_Schmutzer.jpg; p. 13 canbedone/Shutterstock.com; pp. 15, 17 Aphelleon/Shutterstock.com; p.15 https://commons.wikimedia.org/wiki/File:Buzz_Aldrin_and_the_U.S._Flag_on_the_Moon_(9460188482).jpg; p. 16 buradaki/Shutterstock.com; p. 18 Triff/Shutterstock.com; p. 19 Inferiorz Presents/Shutterstock.com; p. 20 Prostock-studio/Shutterstock.com; p. 21 Dotted Yeti/Shutterstock.com; p. 23 Castleski/Shutterstock.com; p. 23 Sandro Pavlov/Shutterstock.com; p. 24 3000ad/Shutterstock.com; p. 25 Bob Pool/Shutterstock.com; p. 26 Gleb Tarro/Shutterstock.com; p. 27 Josephine Julian/Shutterstock.com; p. 28 Lukas Gojda/Shutterstock.com; p. 29 patpitchaya/Shutterstock.com.

Cataloging-in-Publication Data

Names: Harts, Marie.
Title: Gravity / Marie Harts.
Description: New York : Britannica Educational Publishing, in Association with Rosen Educational Services. 2024. | Series: Discover more: exploring physical science | Includes glossary and index.
Identifiers: ISBN 9781641900515 (library bound) | ISBN 9781641900508 (pbk) | ISBN 9781641900522 (ebook)
Subjects: LCSH: Gravity--Juvenile literature.
Classification: LCC QC178.H37 2024 | DDC 531′.14--dc23

Manufactured in the United States of America

CPSIA Compliance Information: Batch #CWBRIT24. For further information contact Rosen Publishing at 1-800-237-9932.

Contents

Gravity Is Everywhere!

Gravity is a force you can find everywhere on Earth and even in space. Because of gravity, all objects in the universe attract all other objects. This force is what keeps everything there is from flying apart! You see gravity at work every day. Animals, plants, buildings, and even air are all affected by gravity. If you drop a book, gravity is what pulls it to the ground. Even your body has gravity! That's because you have mass.

A basketball falls to the ground, even if nothing hits it while it was in the air, due to gravity.

Mass is the amount of matter or "stuff" an object has. Space objects with mass are attracted to other objects with mass due to gravity. Smaller objects with less mass orbit larger objects with more mass.

Gravity is very important in outer space. All the planets **orbit** the sun thanks to gravity. It also keeps the moon orbiting Earth. Gravity pulls everything together like a magnet.

WORD WISE

The path a space object takes around another space object is called an orbit. All orbits are elliptical. This means they are shaped like an ellipse, which is similar to an oval.

The Pull of Gravity

Imagine throwing a small ball as high as you can up into the clear blue sky. No matter how high you throw it, it will eventually fall back to Earth. The force that pulls it back is gravity. Earth's gravity keeps you, and everything else, from flying off into space. Gravity causes any two objects in the universe to be attracted to each other. This attraction is called a gravitational pull. The more mass an object has, the greater its gravitational pull.

It can be fun to jump on a trampoline! You may feel weightless as you are launched into the air by the springs, but gravity will bring you back down.

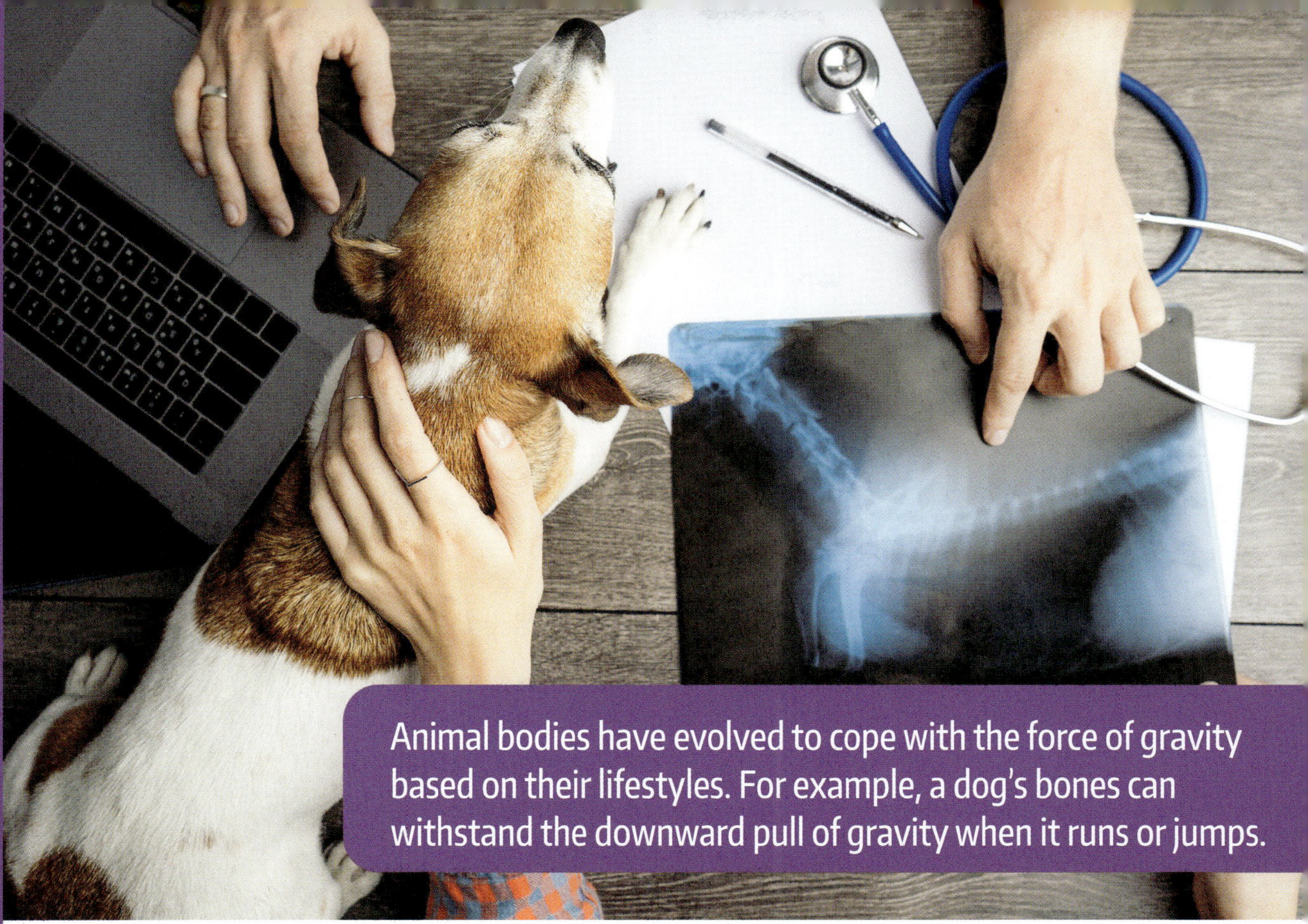

Animal bodies have evolved to cope with the force of gravity based on their lifestyles. For example, a dog's bones can withstand the downward pull of gravity when it runs or jumps.

Gravity is what gives an object weight. When you step on a scale, gravity pulls you against the scale. Your weight on this scale is how much gravity is acting upon you.

Consider This

Gravity has existed throughout Earth's history. This means many plants and animals have evolved to live with gravity. Can you think of some examples of their features?

A Law from a Falling Apple

More than 300 years ago, an English scientist named Isaac Newton made a discovery in a surprising way. The story goes that Newton was sitting in a garden when he saw an apple fall from a tree. He thought about what made the apple fall straight to the ground. Newton knew there must be a force that acted on the apple and that understanding this force could explain other things, such as why the moon stays in orbit around Earth.

Isaac Newton put forward his law of universal gravitation in 1687. He believed it explained the movement of planets and their orbiting moons.

By dropping objects of different weights from the Leaning Tower of Pisa, Galileo Galilei legendarily showed that objects of different masses fall with the same acceleration.

Newton wrote the law of universal gravitation. According to this law, any two objects in the universe with mass have a gravitational force between them.

Consider This

Italian astronomer Galileo Galilei might have been the first to suggest that the planets' movements and falling objects are both controlled by the same force. However, Newton was the first to use an exact mathematical form to define gravity. Why do think Newton's discovery was so important?

The Magic of Mass

Based on Newton's idea, people can predict, or guess, what gravity will do based on two features. The first is mass. The bigger an object is, or often the more mass it has, the stronger its gravitational pull is. Earth is very large. So, when you are standing next to a house, you are pulled toward Earth and not the house because Earth is much bigger than a house.

Gravity in space pulls dust and gas into what are called interstellar clouds.

Saturn has a greater gravitational pull than Earth. Its surface gravity is about 107 percent of Earth's surface gravity. So, you would weigh 7 pounds (3.2 kg) more on Saturn than you do on Earth!

Because people know that all objects with mass have gravity, they know that even a tiny speck of dust has a gravitational pull. But its gravity is very weak. Even though it has weak gravity and can be easily carried by air currents, eventually, dust always settles due to Earth's gravitational pull.

Consider This

How large does something such as a planet have to be for something else to orbit it? A hint is that there is no specific or "magic" mass or size.

Dependable Distance

What's the second feature that Newton said affects gravity? It's distance! The farther apart two objects are, the weaker the gravitational pull is between them. If Earth were as far away from the sun as Mars, the gravitational pull of the sun would be weaker because it would be farther away.

Much of what we know today about gravity comes from Newton's law of gravitation and German-born physicist Albert Einstein's theory of general relativity.

According to Einstein's theory of general relativity, the larger an object is, the more its gravity can curve and warp, or bend, the space around it, like a ball resting on a heavy sheet. This warping also affects time.

The smaller the distance between two objects, the stronger the force of gravity is between them. There is a star in the universe that is 315 **solar masses**! Even though it's bigger than our sun, the planets in our solar system are attracted to our sun because it's much closer.

WORD WISE

One solar mass is equal to the mass of the sun.

The Moon's Marvelous Path

The moon is pulled towards Earth's gravity because Earth is much bigger. If Earth had no gravity, the moon would travel in a straight line. But Earth's gravity is always pulling the moon toward Earth. As a result, the moon travels in a continuous path around Earth.

The moon never stops moving in its orbit. This is thanks to the Earth's continuous gravitational pull. If the moon did stop orbiting, it would crash into Earth! Although the moon is continually falling towards Earth's gravity, it has just the right amount of speed and gravity to fall around Earth instead of into it.

Astronauts' movements on the moon look bouncier due to the moon's lack of gravity compared to Earth.

compare and contrast

Both the Earth and the moon have a gravitational pull. However, because it is larger, the Earth's gravitational pull is greater. How do you think this affects your weight on the moon?

Density is how much mass is found in a given space. The moon is only about 60 percent as dense as Earth.

Orbiting the Sun

Earth is one of eight planets in our solar system. There are also comets, asteroids, moons, and a lot more. Because of gravity, all the planets in the solar system orbit the sun just as the moon orbits Earth.

The closer a planet is to the sun, the faster its orbit. Mercury is the closest planet to the sun, so it has the fastest orbit around the sun at 88 Earth days.

Neptune is the farthest planet from the sun. It also has the slowest orbit around the sun at about 165 Earth years.

The sun is a star and much bigger than all the planets put together. So even though the planets are trying to fly away from the sun, the sun is pulling them toward it. This effect keeps them in orbit, moving around the sun for billions of years. When a planet moves closer to the sun, its orbit speeds up. As it moves away from the sun, its orbit slows down.

Consider This

Many comets also orbit the sun. While the planets' orbits are almost circular, comets' orbits are more elliptical or "squashed" and can take them beyond the solar system. Why do you think comets have more elliptical orbits?

Keeping Everything Together

Our solar system is just one of 3,200 planetary systems in our galaxy, called the Milky Way. A galaxy is made up of solar systems, clouds of gas, and dust particles that move together through the universe.

Black holes are formed from the remains of a large star that has died in a huge explosion called a supernova.

The Milky Way is a fairly thin, flattened disk. From Earth, we are looking in the direction of the disk, so the Milky Way looks like a band in the sky.

Gravity keeps everything in the Milky Way galaxy together. On average, the moon is 239,000 miles (384,600 km) away from Earth, but Earth's gravity is strong enough to keep the moon in orbit. Every moon orbits a planet. Every planet orbits a star. Even stars orbit the center of a galaxy. At the center of most galaxies is a black hole. However, everything in the galaxy orbits the galaxy's center of mass, not the black hole. Although black holes are big, the one at the center of the Milky Way is only about one millionth of our galaxy's total mass.

WORD WISE

A black hole is a place in space where a lot of matter has been squeezed into a tiny space. This causes a gravitational pull that is so strong, not even light can escape.

Super Satellites

A satellite is any object that orbits a planet. Earth has a natural satellite—the moon. Besides the moon, many human-made satellites orbit Earth. These satellites are used to take pictures, help make phone calls, and relay TV shows, all from outer space!

Homes in the mountains or countryside may get their internet from satellites as opposed to cables.

There are around 2,000 active satellites orbiting Earth, along with 3,000 dead satellites.

When a satellite is sent into space, Earth's gravity causes it to orbit the planet, just like the moon. Satellites can continue to orbit Earth even after they stop working. If the satellite is closer to Earth, it will use its last bit of fuel to slow down and fall out of orbit. If it is father from Earth, it will be sent even farther away.

Consider This

Dead satellites are among the 34,000 pieces of "space junk" larger than 3.9 inches (10 cm) orbiting Earth. What do you think are the dangers of space junk such as this, and how can it be cleaned up?

Why Astronauts "Float"

Astronauts must train to live in space, where they experience a feeling of weightlessness. But are they *really* weightless? Gravity is pulling on them, just as it does the rest of us. But they are moving sideways around Earth in an orbit.

From the point of view of astronauts in orbit, there's a force that seems to be pulling them away from Earth at the same time that Earth's gravitational pull is pulling them toward the center of the planet with equal strength. This apparent force, called centrifugal force, makes it look as if the astronauts are floating in space. A centrifugal force acts on any object that is moving in a circular path. It's the feeling of "flying off" you may have when riding a merry-go-round or in a car going around a turn.

When in space, astronauts experience what is called microgravity. This is the condition where gravity seems very small.

compare and contrast

Astronauts often train underwater because it allows them to practice feeling weightless. What other ways do you think being underwater is similar to and different from being in space?

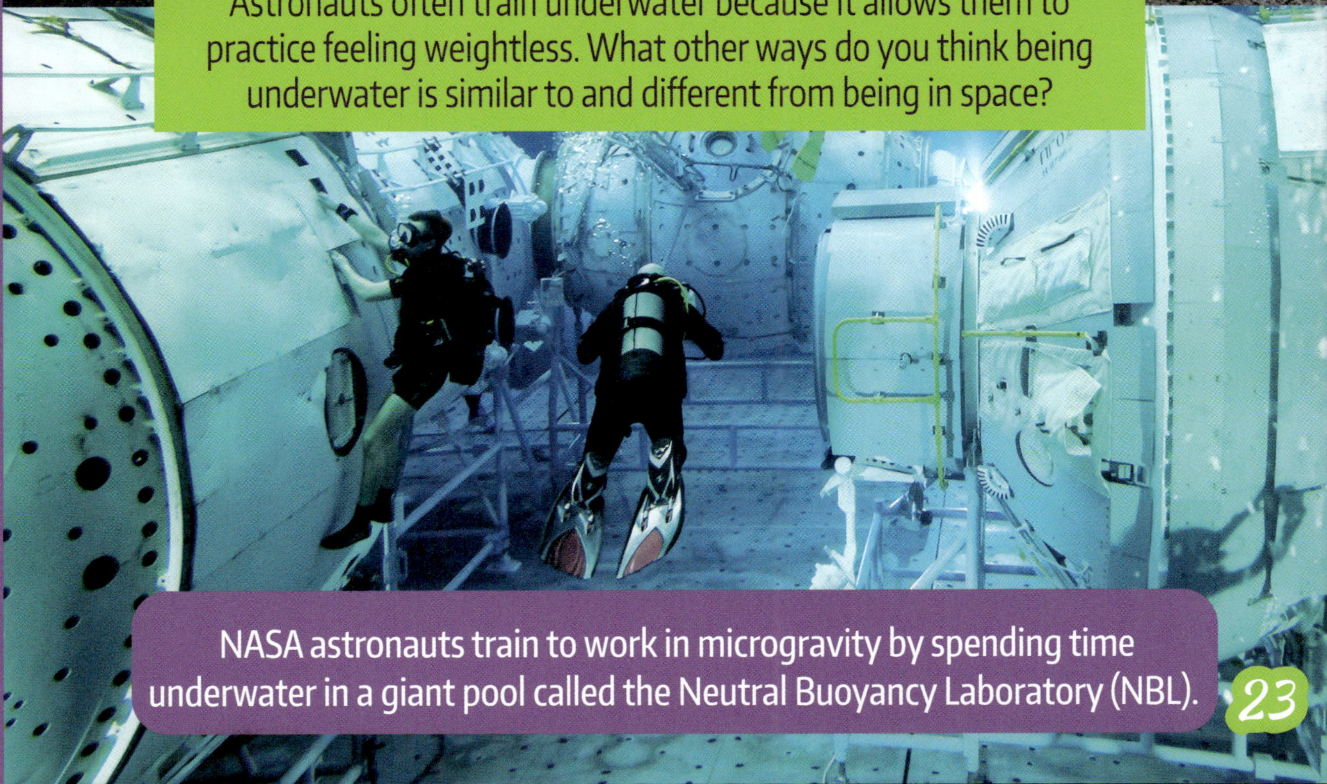

NASA astronauts train to work in microgravity by spending time underwater in a giant pool called the Neutral Buoyancy Laboratory (NBL).

Terrific Tides

Although it's smaller than Earth, the moon's gravity affects Earth. The moon's gravity pulls on Earth and its oceans, causing a regular pattern of tides as Earth rotates each day. On the side of Earth near the moon, the moon's gravity causes the ocean to bulge upward. This causes high tides along beaches. The ocean also bulges on the opposite side of Earth. This happens because the moon is pulling Earth away from the ocean on that side.

The moon takes about 27 days, or one month, to complete its orbit around Earth.

Tide pools are areas that fill with water as a tide recedes. While they may seem small, they are home to many different plant and animal species, or kinds of creatures.

The tides change as the moon moves around Earth. About once a month, the moon is closer to Earth. This is called perigee, and it means there is a greater range in Earth's high and low tides.

Consider This

Tidal movements help keep Earth's climate stable and churn up material that underwater sea creatures need to survive. What do you think would happen to life on Earth if the moon suddenly disappeared?

Shaping Earth's Surface

Have you ever wondered how land formations such as the Grand **Canyon** came to be? They are formed due to erosion. This is when rocks and earth are worn away by natural forces, such as flowing water. After the rocks and earth start to wear away, they are moved to a new place. Erosion has helped shape Earth's surface by making features such as mountains and valleys.

Over millions of years, the Colorado River carved what is today the Grand Canyon.

A landslide is one example of what's called gravity erosion or mass movement. This is when land and soil move down a slope due to the pull of gravity without the involvement of water or ice.

Water, wind, and ice are the main things that cause erosion. Gravity can encourage erosion too. For example, dirt and rocks on the side of a mountain can be loosened by rain and wind. Over time, a landslide can happen when gravity pulls the dirt and rocks down the mountain.

Consider This

Planting trees is a way to prevent erosion because the tree roots can grab onto soil. Can you think of other ways people can prevent erosion?

WORD WISE

A canyon is a deep valley with steep sides created by erosion over millions of years.

A Force with Unknowns

Much is still unknown about gravity. Isaac Newton wasn't sure why gravity exists, but his law is a theory that explains what gravity does. Even though people don't know why gravity works, they see it working every day.

Despite weighing up to around 220 tons (199.6 mt), planes can fly because their wings are shaped to create enough lift to overcome gravity.

Did you know you can defy gravity? You simply need to lift something in the air! Gravity is the weakest of the four fundamental forces of nature. The other forces are electromagnetism and a weak and strong nuclear force.

Without gravity, when you jump in the air, you would never come down! Our planet would have no moon. Earth and the other planets in our solar system would not be able to travel around the sun. They might even run into each other! If gravity didn't exist, our solar system and even the universe would not be what they are today.

Consider This

Microgravity in space can cause health problems for astronauts such as muscle atrophy. This is when muscles start thinning and wasting away due to lack of use. Can you think of any ways to prevent this?

Glossary

air resistance: The force that acts in opposition to the motion of an object as it moves through the air, slowing the object down.

astronomer: A person who studies stars, planets, and other heavenly bodies.

buoyancy: An object's ability to float in water or air.

climate: The average weather conditions of a place over a period of time.

comet: A space object made of ice and dust that has a long glowing tail when it passes close to the sun.

electromagnetism: A magnetic field that is made by a current of electricity.

evolve: To grow and change over time.

launch: To send out with great force.

legendarily: Of, described in, or based on a legend, which is a story from a past event that many people believe, but cannot prove.

muscle: One of the parts of the body that allow movement.

nuclear: Having to do with the power created by splitting or joining together the nuclei, or cores, of atoms. Atoms are the smallest bits of matter.

physicist: A scientist trained in physics, which is the study of nature, especially the behavior of matter and energy.

recede: A word to describe something that becomes further and further away.

satellite: An object in space that moves around another object.

solar system: The sun and everything that travels around it.

theory: An idea that is intended to explain facts.

tide: The regular rising and falling of ocean levels.

trampoline: A tough sheet often made of a material called nylon that is supported by springs on a metal frame. It is often used as a springboard for jumping and landing.

universe: All existing matter and space.

For More Information

Books

Hansen, Grace. *Gravity*. Minneapolis, MN: Abdo Kids Jumbo: 2019.

Pettiford, Rebecca. *Gravity*. Minneapolis, MN: Bellwether Media, 2019.

Wood, Alix. *Sir Isaac Newton*. New York, NY: PowerKids Press, 2019.

Websites

Gravity
kids.britannica.com/kids/article/gravity/400109
Find more information about gravity here, including a diagram and video.

Science Trek: Gravity!
www.pbs.org/video/gravity-m3swlv/
Watch a fun video that explains more about how gravity keeps everything on Earth from flying off into space.

What Is Gravity?
spaceplace.nasa.gov/what-is-gravity/en/
Learn more about gravity and its importance in our solar system here.

Index

A

C

D

E

G

M

N

O

P

S

W